Bricoleur's Algebra

Tarek Halabi

2026

SLINGS AND ARROWS PRESS

First edition, 2026

Published by Slings and Arrows Press

Contents

If you have ever felt frustrated by how unceremoniously groups are sometimes introduced, their definition simply springing into your face with no motivation beyond the assurance that "they are everywhere";
if you have felt bewildered when a teacher walks in one day and decrees that we are now working with two binary operations instead of one, and that one must "distribute" over the other—again with no motivation beyond the observation that this happens for integers, matrices, and other examples we already know; and if it saddens you that this makes abstract algebra difficult to defend against the accusation that it is a song and dance designed around things we all already knew;
if you can follow the mathematics on the blackboard, perhaps even earn high marks on the exams, yet still have no idea what big-picture problem the subject is trying to solve;
if you have ever worried that the lecturer was committing a kind of tautological conceit by skipping the proof that matrix multiplication is associative because matrix multiplication represents composition of linear maps, after spending five seconds waving a hand at the associativity of composition itself; and if you are a visual learner who would prefer actually to see associativity rather than watch someone wave a hand at it:

Then this book is for you.

Why this definition and not another? Who decides what we spend our time on? And how do we defend against the accusation that abstract algebra is a song and dance designed—through otherwise unmotivated definitions—around things we already know, like arithmetic and linear algebra?

A definition is, by definition, a construction going from general to specific. We need to place on this construction a filter that we can all agree on. A familiar one already exists: canonical construction. Canonical means making no arbitrary choices within the construction once the components are given. If the goal is to be specific yet canonical, we can go a step further.

Metacanonical construction adds the criterion that the selection of which canonical constructions matter must itself be non-arbitrary. A familiar device offers one implementation: mirrors. We canonically pick out canonical constructions that lie entirely within the mirrored space of a mirror—that is, a correspondence. As it turns out, the objects produced by such constructions are precisely the ones that stand out most in importance.

The organizing principle of algebra is then to move from the general to the specific without making any arbitrary choices. Here, we demonstrate this approach first for binary operations, then associativity, and finally groups.

Some of the visual conventions used by cards throughout this text are:

Definition carved out of upstream definitions or stipulations.

Stipulation: definition carved out of thin air.

Result.

Canonical construction. These may include definitions and results.

Meaning.

Chapter 1

Maps

1.1 Sets

An element is a something.

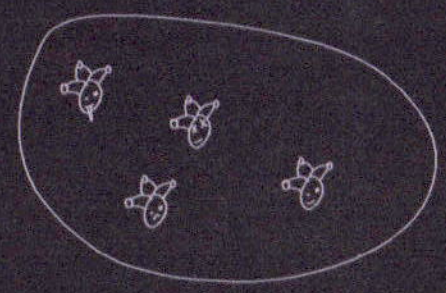

A set is a collection of distinct elements.

Stipulations are typically followed by a series of definitions. I will try to list the definitions left to right, top to bottom in increasing order of "specified." On the one hand, we are more likely—or more easily able—to find canonical constructions of definitions that are more specified. On the other hand, canonical constructions of more general definitions are more likely to serve as raw material for other constructions. For this reason, we want to make as many definitions as possible while covering as much of the spectrum from general to specified as we can.

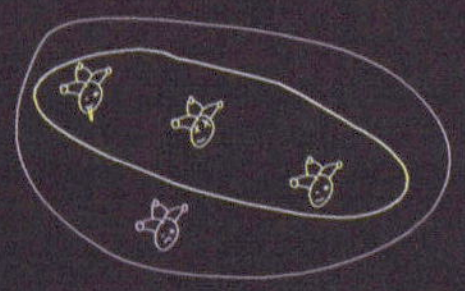

A **subset** A of B is a set whose elements also belong to B.

Do not let the figure fool you into thinking that we already have a canonical construction of subsets. We have not stipulated jesters, and so we have not stipulated happy or sad jesters. The figure is for illustration purposes only. We have stipulated that we have collections of somethings and the reader might conclude that we can always choose some of a collection of somethings. This would certainly be a construction, but not a canonical one. Canonical means "no arbitrary choices beyond what is given".

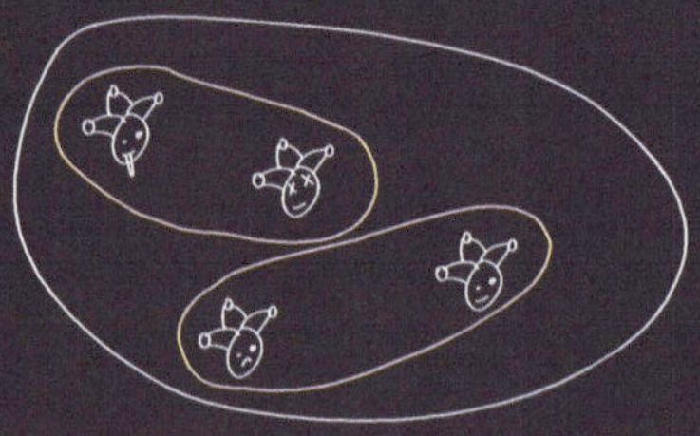

A **partition** P of a set A is a set of nonempty subsets of A such that every element of A belongs to exactly one element of P.

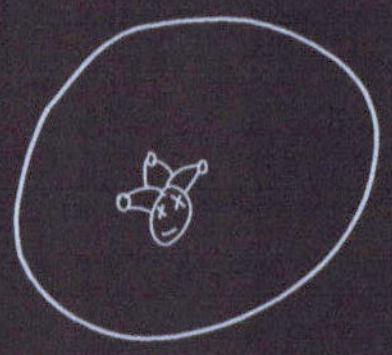

A **singleton** is a set containing exactly one element.

Our first canonical constructions are all trivial.

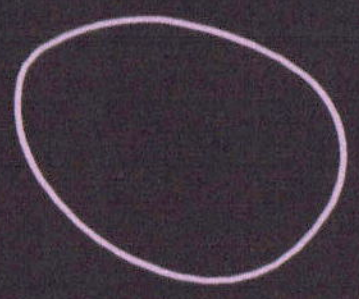

The **empty set** is a canonical construction of a set.

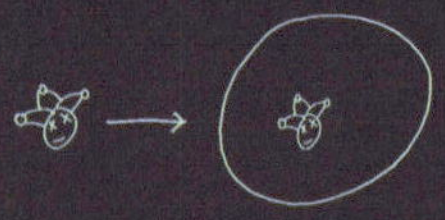

Given an element, we can use it to canonically construct a **singleton**.

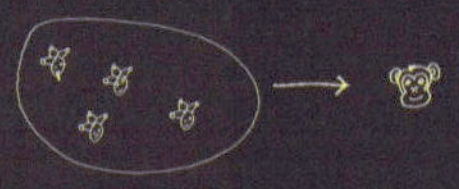

Given a set, we can use it to canonically construct an **element**.

Given sets $A, B, \ldots$, taken in that order, we can canonically construct their **Cartesian product** $A \times B \times \ldots$, whose elements are the tuples $(a, b, \ldots)$ with $a \in A$, $b \in B$, and so on.

Given a set of sets $\{A, B, \ldots\}$, we can canonically construct their **union** $A \cup B \cup \ldots$, containing the elements that belong to at least one of these sets.

Given a set of sets $\{A, B, \ldots\}$, we can canonically construct their **intersection** $A \cap B \cap \ldots$, containing the elements that belong to every one of these sets.

1.2 Maps

A map is a magical tube with a set of allowed inputs, called its domain, and a set in which its outputs must land, called its codomain. Feed it any element of its domain, and it produces exactly

A **partial self-map** is a map whose domain is a subset of its codomain.

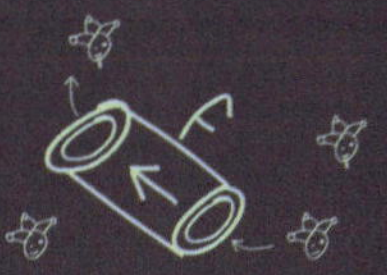

A **self map** is a map whose domain is the same as its codomain.

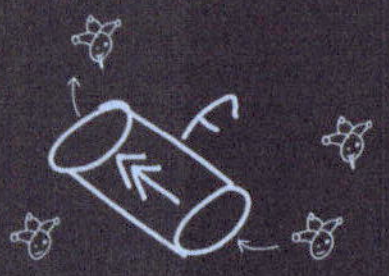

A **surjective map** is a map that outputs every element in its codomain for at least one element in its domain.

An **injective map** is a map that never outputs the same output for two different inputs.

A **bijective map** is a map that is both surjective and injective.

A **permutation** is a map that is both a self map and a bijection.

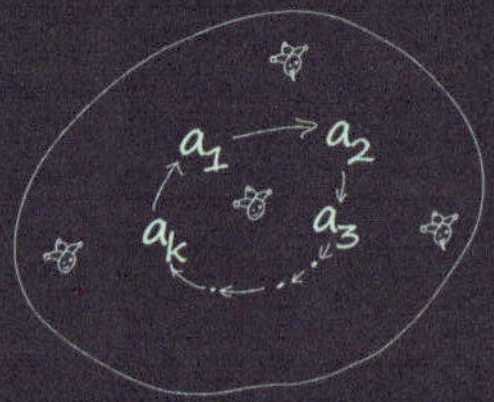

A k**-cycle** is a permutation that, for some distinct elements $a_1, \ldots, a_k$, maps a_1 to a_2, a_2 to a_3, ..., a_{k-1} to a_k, a_k to a_1, and maps every other element to itself.

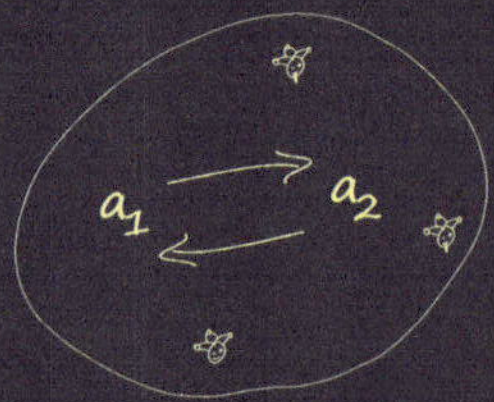

A **transposition** is a 2-cycle map.

Definition 1.1 The **identity map** maps every element to itself.

Definition 1.2 A map f is **composable** with a map g if the codomain of g is the domain of g.

The number of elements of a set A is called its **order** $|A|$.

The **identity map** is a canonical construction of a map.

Given a 2-element subset, we can canonically construct the corresponding **transposition map**.

Given an ordered k-tuple of distinct elements, we can canonically construct the corresponding k**-cycle map**.

Given a map, we can canonically construct a subset of its codomain, called its **image**, containing all elements that are outputs of the map for at least one element of its domain.

Given a map f, and an element y from f's codomain, we can canonically construct a subset of its domain, called f's **fiber** over y, containing all elements that f maps to y.

Given a map f, we can canonically construct a partition of its domain, called f's **fiber partition**, given by the set of all of f's non-empty fibers.

Given a self map f, and an element x in f's domain, we can canonically construct a subset of f's domain called f's **orbit** of x, which is the set of elements reached by repeatedly applying f forward (and in the case of permutations backward as well).

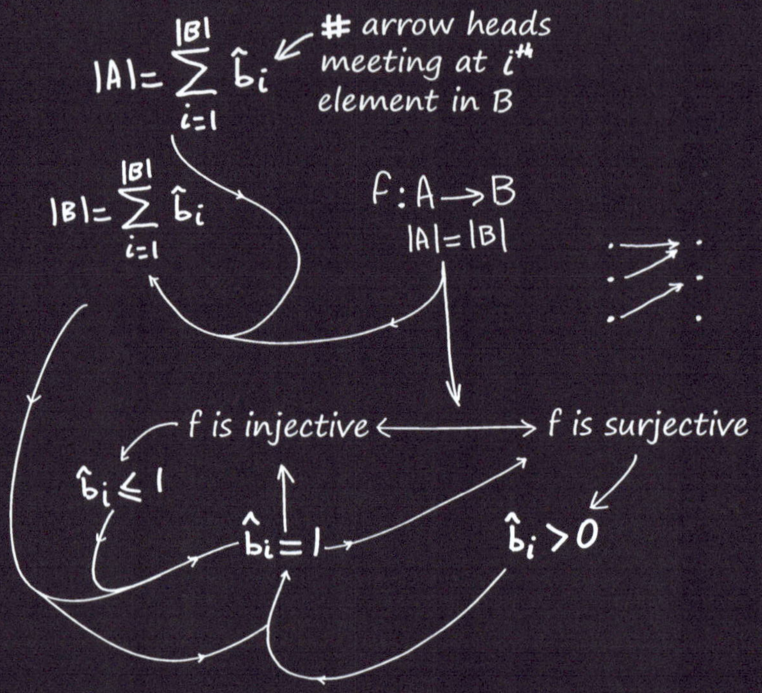

★ **Result 1.1** Let $f\colon A \to B$ be a map between two finite sets. If $|A| = |B|$, then f is injective if and only if f is surjective.

Given a self map of a finite set that is injective, Result 1.1 shows that it is surjective, and hence a **permutation**.

Given a self map of a finite set that is surjective, Result 1.1 shows that it is injective, and hence a **permutation**.

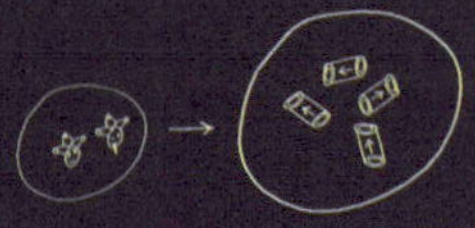

Construction 1.1 Given a set S, we can canonically construct the **set** $\{S \to S\}$ **of all self maps of** S.

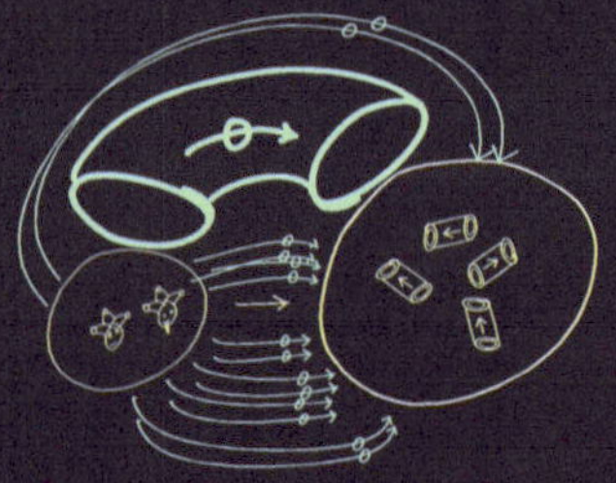

Construction 1.2 Given a set S, we can canonically construct the set $\{S \to \{S \to S\}\}$ of all maps from S to the set of all self maps of S. We will colorfully refer to these maps as **comets**.

Given a set A, we can canonically construct the **set** Sym(A) **of all permutations of** A, also trivially a subset of the set of self maps of A.

1.3 Binary maps

For a positive integer n, an **n-ary map** is a map whose domain is a subset of the Cartesian product $A_1 \times \cdots \times A_n$ of n sets, called the factors of the domain.

A **ternary map** is a 3-ary map; equivalently, its domain is a subset of the Cartesian product $A_1 \times A_2 \times A_3$.

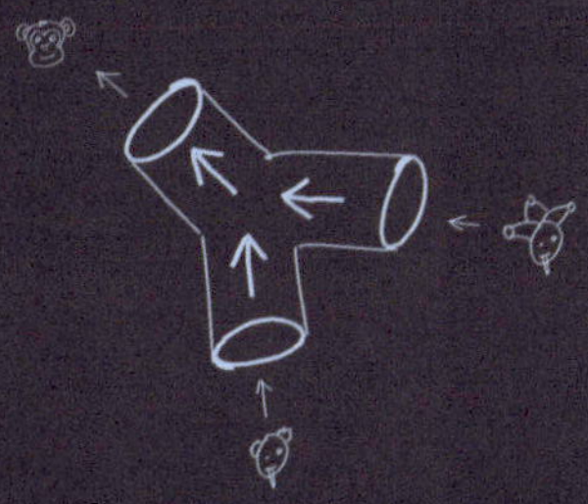

A **binary map** is a 2-ary map; equivalently, its domain is a subset of the Cartesian product $A_1 \times A_2$.

At this point, the pattern may be clear: each canonical construction puts something new on the workbench, where it becomes raw material for the next. Abstract algebra is a song and dance after all—the self-generating song and dance of canonical construction.

The self-generation demonstrated in this text, however, immunizes abstract algebra against the accusation that it is built around "things we already know," such as integers and matrix multiplication. This is the key idea of the book: a shift to the deeper and more accurate perspective that the things we already know are echoes of the self-generating song and dance of canonical construction.

The construction that follows, Machine 1.1, perfectly demonstrates this self-generated momentum. We begin with a pair of maps from S to O,

$$\text{dom}, \text{cod}\colon S \to O.$$

They lead us to $S \times S$, where they canonically construct the subset D of composable pairs. They also canonically construct the typed sets, whose nonempty members partition S. Looking back in the opposite direction at $\{D \to S\}$, the set of maps from D to S, we can now use the typed sets to canonically construct the subset of $\{D \to S\}$ consisting of maps that match those types. The elements of this final subset are precisely the binary multiplications. Each canonical construction has supplied the raw material for the next.

Machine 1.1 – Binary multiplication

Given two maps,

$$\mathsf{dom}, \mathsf{cod}\colon S \to \mathcal{O},$$

where $\mathcal{O}$ is called the set of labels, we can canonically construct:

Construction 1.3 The **subset**

$$D = \{(a,b) \in S \times S : \mathsf{dom}(a) = \mathsf{cod}(b)\} \subseteq S \times S,$$

whose elements are called **composable pairs**.

The fibers of the map:

$$T\colon S \to \mathcal{O} \times \mathcal{O}, \qquad T(a) = (\mathsf{dom}(a), \mathsf{cod}(a))$$

called the **typed sets**:

$$S_{X,Y} = \{a \in S : \mathsf{dom}(a) = X,\ \mathsf{cod}(a) = Y\}$$

Construction 1.4 From D and the typed sets, we can canonically construct the **subset of** $\{D \to S\}$ consisting of all maps m such that, for every $(a,b) \in D$,

$$m(a,b) \in S_{\mathsf{dom}(b),\mathsf{cod}(a)}.$$

Equivalently,

$$\mathsf{dom}(m(a,b)) = \mathsf{dom}(b), \qquad \mathsf{cod}(m(a,b)) = \mathsf{cod}(a).$$

and define:

Definition 1.3 An element of the set in Construction 1.4 is called a **binary multiplication**.

Construction 1.5 Given a set S of maps between the sets in a set $\mathcal{O}$, we trivially have the canonical **set of composable pairs** from Construction 1.3: the ordered pairs $(a, b) \in S \times S$ for which b is composable with a in the sense of Definition 1.2. This is a specialization of the composable-pair construction in Machine 1.1.

Definition 1.4 A **binary operation** is a binary multiplication whose set of labels $\mathcal{O}$ is a singleton. It has only one typed set which is called a **magma** under this binary operation.

Construction 1.6 Given a nonempty set S, we can canonically construct the **set of all magmas on** S by taking the set $\{S \times S \to S\}$ of all maps from $S \times S$ to S. By Definition Definition 1.4, these maps are precisely the binary operations on S.

If a binary operation on M is also a binary operation on a subset S of M, then S is called a **submagma** of M.

Given a set $\mathcal{O}$ of sets, and a set of maps S between the sets in $\mathcal{O}$ such that the composite of every composable pair in S also belongs to S, **composition** of these maps as shown in the figure is a canonical construction of a binary multiplication.

Chapter 2

Associativity

With a Surprise Appearance by Distributivity

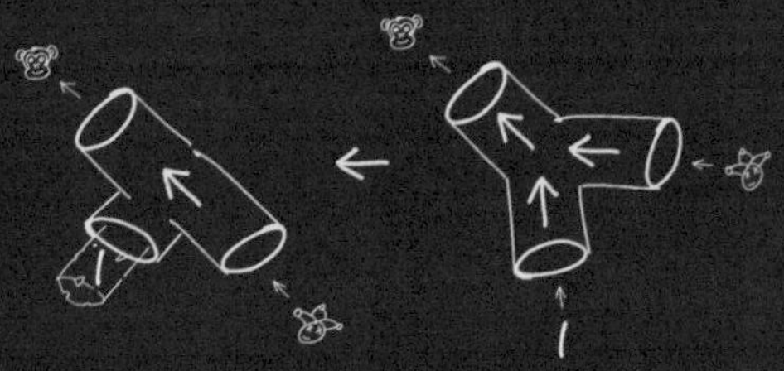

Construction 2.1 *Given a binary map*

$$f\colon D \to C, \qquad D \subseteq L \times R,$$

and an element $l \in L$ we can canonically construct, as shown in the figure, a map from $\{r \in R : (l, r) \in D\}$ to C, called ***left multiplication by l under f.***

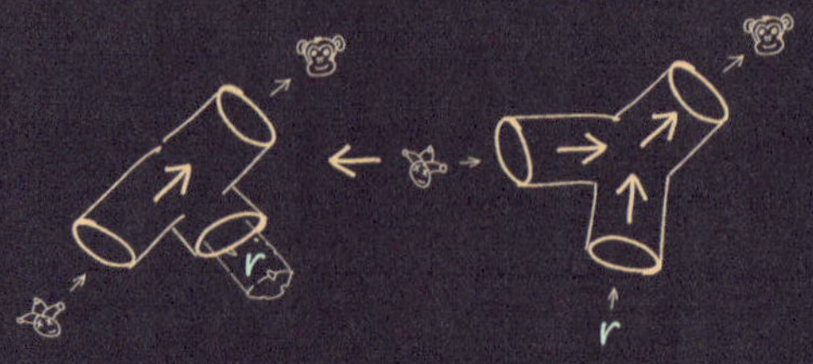

Construction 2.2 Given a binary map

$$f\colon D \to C, \qquad D \subseteq L \times R,$$

and an element $r \in R$ we can canonically construct, as shown in the figure, a map from $\{l \in L : (l,r) \in D\}$ to C, called **right multiplication by r under f.**

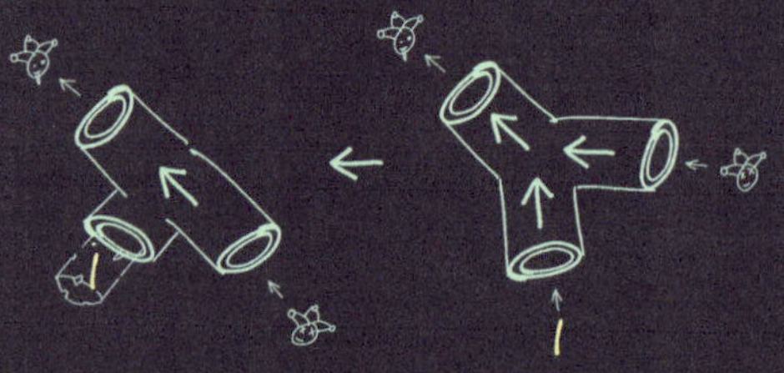

Construction 2.3 Given an element l of a magma M, we can canonically construct, as shown in the figure, a self map $M \to M$ called **left multiplication by l under M's binary operation.**

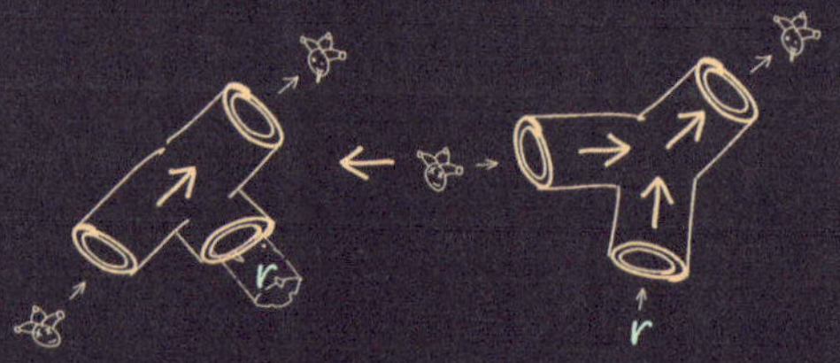

Construction 2.4 Given an element r of a magma M, we can canonically construct, as shown in the figure, a self map $M \to M$ called **right multiplication by r under M's binary operation.**

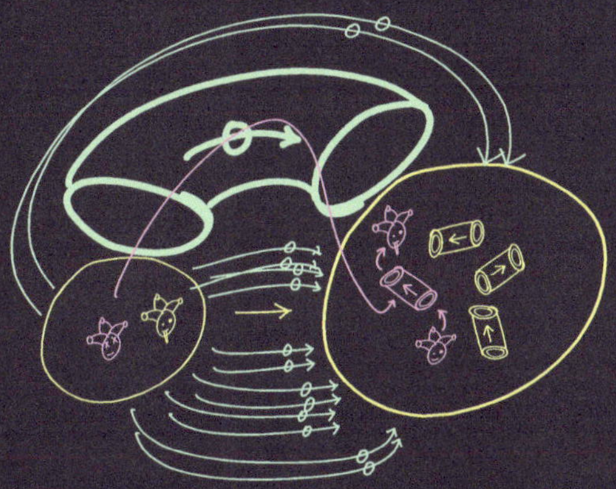

Construction 2.5 Given a comet (see Construction 1.2), we can canonically construct a **magma on** S as follows. For any left input from S use the given comet to pick out a self map s on S. On any right input from S use s to obtain the output of our canonically constructed binary operation on S.

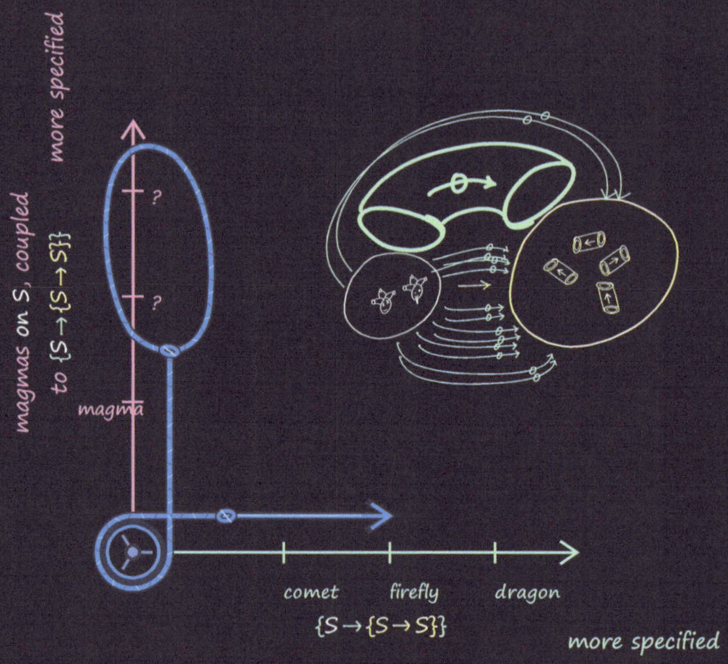

Construction 2.6 Given a magma M on S, repeat Construction 2.3 for each element of S to canonically construct a map from S to the set of all self maps of S (see Construction 1.2). We will colorfully refer to this as M's **comet**. It is more commonly known as M's **left multiplication map**, but that name unnecessarily increases the chances it will be conflated with its similarly named outputs (the latter named in Construction 2.3). It also follows from Construction 2.5 that given a map from S to the set of all self maps of S, we can canonically construct a magma on S. This relationship is represented in the figure above as coupling between an arrow and a lasso.

Meaning 2.1 It follows from Construction 2.6 that the set of all comets of S shown in Construction 1.2 means the set of all magmas on S.

Construction 2.6 is a mirror between the axes shown in the figure: magmas on S, and comets. It presents an opportunity to move further from the general towards the specific without making arbitrary choices: we canonically pick out canonical constructions that are entirely within its mirrored space, if there are any. The gambit is as follows: construct along the two axes joined by the correspondence and find that, once the second construction is carried back across the correspondence, both constructions reach the same subset of magmas.

We've already completed one iteration of the above described gambit. The set of all magmas on S is a trivial subset of itself. It was canonically constructed in Construction 1.6 directly on the magma side of our mirror, Construction 2.6. We had already canonically constructed the set of all comets in Construction 1.2 on the other side of the same mirror. The mirror, Construction 2.6, itself showed that when the set of all comets is carried across to the other side, it coincides precisely with the set of all magmas on S. The set of all magmas on S is therefore crowned as an object produced by a **metacanonical construction**: a canonical construction canonically selected because it lies entirely within the mirrored space of a mirror (correspondence).

We next make a second pass through the gambit, this time seeking a proper subset of the set of all magmas on S. Throughout the text, we'll call construction on the magma axis Lasso, and construction on the comet axis Arrow. Lasso evokes the notion of tightening from general, or "loose," definitions to more specific, or "tightened," definitions. Arrow evokes the notion of piercing through to reach deeper meaning.

2.1 Lasso

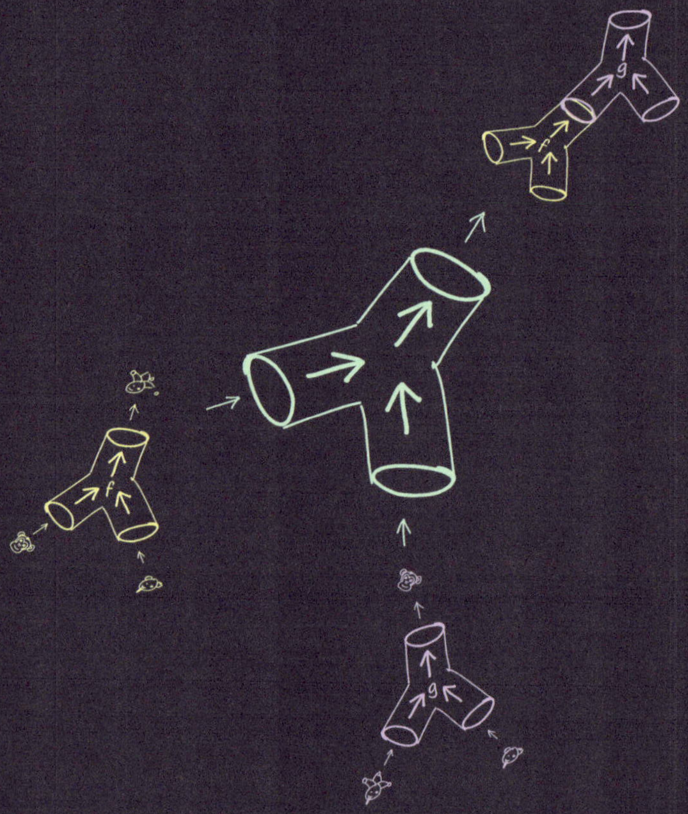

Given binary maps

$$f\colon D_f \to S, \qquad g\colon D_g \to S, \qquad D_f, D_g \subseteq S \times S,$$

we can canonically construct the ternary map called their **left composition**:

$$L_{f,g}\colon D_L^{f,g} \to S, \qquad L_{f,g}(a,b,c) = g\big(f(a,b),c\big),$$

where

$$D_L^{f,g} = \big\{(a,b,c) \in S^3 : (a,b) \in D_f,\ \big(f(a,b),c\big) \in D_g\big\}.$$

Thus f is first applied to a, b, and then g is applied to that output and c.

Given binary maps

$$f\colon D_f \to S, \qquad g\colon D_g \to S, \qquad D_f, D_g \subseteq S \times S,$$

we can canonically construct the ternary map called their **right composition**:

$$R_{f,g}\colon D_R^{f,g} \to S, \qquad R_{f,g}(a,b,c) = f\big(a, g(b,c)\big),$$

where

$$D_R^{f,g} = \big\{(a,b,c) \in S^3 : (b,c) \in D_g,\ \big(a, g(b,c)\big) \in D_f\big\}\,.$$

Thus g is first applied to b, c, and then f is applied to a and that output.

Definition 2.1 Given binary maps $f\colon D_f \to S$ and $g\colon D_g \to S$, with $D_f, D_g \subseteq S \times S$, the ordered pair (f, g) is called **associative** if

$$L_{f,g} = R_{f,g}.$$

Because equality of maps includes equality of their domains, this requires $D_L^{f,g} = D_R^{f,g}$ as well as equal outputs on that common domain. When $f = g$, this is the ordinary associativity of the single binary map f.

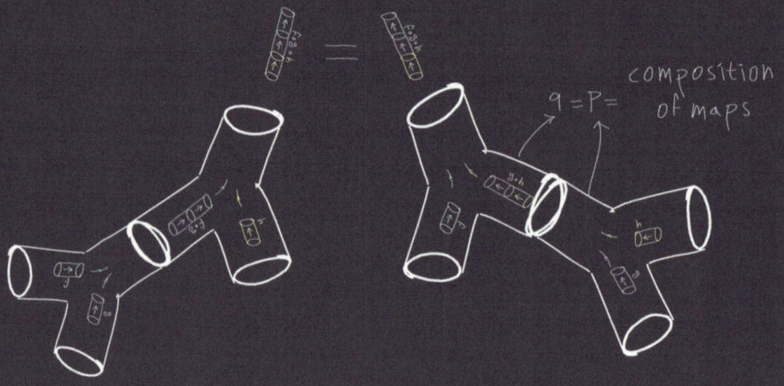

Result 2.1 This figure shows that composition of maps is associative.

Construction 2.7 We can use Definition 2.1 to canonically construct the subset of the set of all magmas known as the set of all **associative magmas or semigroups** on S.

Construction 2.7 concludes the first step in our second gambit for the prize! Next we begin construction on the comet axis, following a path we call Arrow, hoping to find a canonical construction of a subset of the set of all comets on S, in the further hope of finding that when this subset of comets is carried back across the correspondence in Construction 2.6, it coincides with the subset in Construction 2.7.

2.2 Arrow

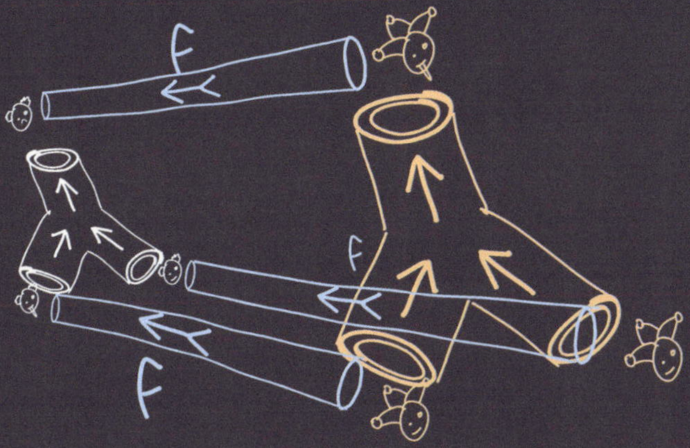

$$f(x) \cdot f(y) = f(x * y)$$

binary operation of second magma

binary operation of first magma

Definition 2.2 A **magma homomorphism** is a map from one magma to another satisfying the condition shown in the figure.

Definition 2.3 A **magma isomorphism** is a bijective magma homomorphism.

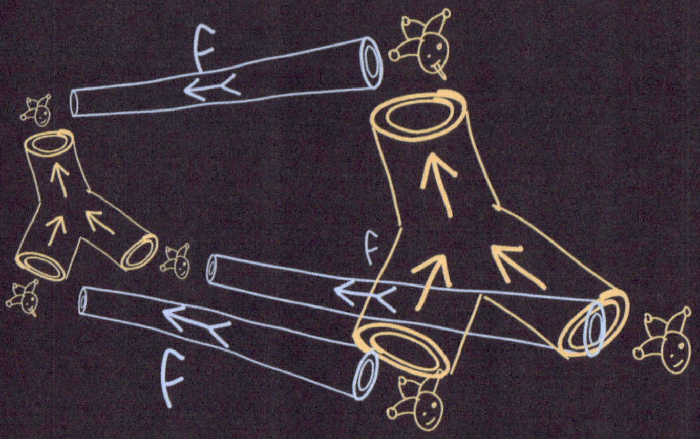

$$f(x) * f(y) = f(x*y)$$

Definition 2.4 A **magma endomorphism** is a magma homomorphism from a magma to itself.

Construction 2.8 Given a magma M on an underlying set S we can canonically construct a subset of $\{S \to S\}$ by taking only those that are magma endomorphisms of M. We will denote this subset as End(M).

Construction 2.9 Given a map, a binary operation on its domain, and a binary operation on its codomain, we can use the equation in Definition 2.2 to canonically construct a **yes or no answer** indicating whether the map is a magma homomorphism.

We cannot yet use Construction 2.9 to claim a canonical construction of a subset of the comets. In order to test whether or not a map from a set S to the set of all self maps of S is a homomorphism we need to specify for each such map, a binary operation (magma) on S and a binary operation (magma) on self maps of S.

Result 2.2 Composition of injections is an injection. Composition of surjections is a surjection. Composition of self maps is a self map. Composition of permutations is a permutation.

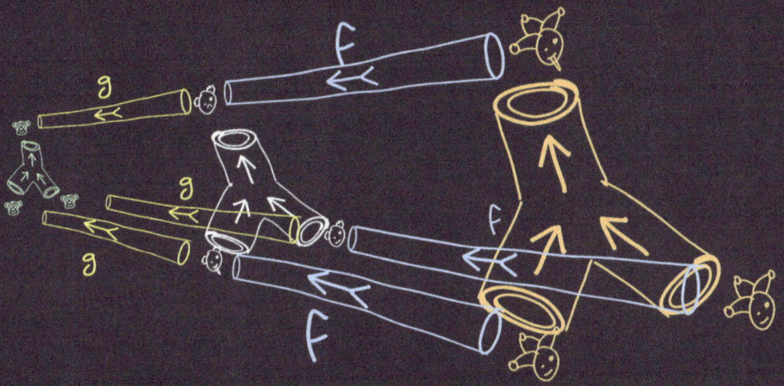

★ **Result 2.3** Composition of magma homomorphisms is a magma homomorphism.

★ **Result 2.4** From Result 2.2 and Result 2.3, composition of magma endomorphisms is a magma endomorphism.

Construction 2.10 It follows from Result 2.2 that we can canonically construct a **magma** by using composition as the binary operation on Construction 1.1. We will denote this magma as $(\{S \to S\}, \circ)$.

Definition 2.5 A comet that is also a magma homomorphism from the comet's magma (see Construction 2.6) to the magma of self-maps under composition is colorfully called a **firefly**.

Construction 2.11 We can now finally canonically construct a **subset of the comets that are fireflies** by taking only the ones that are homomorphisms from the comet's magma (using Construction 2.6) to the magma on $\{S \to S\}$ given by composition.

Not every binary operation on self maps of a fixed set is the composition of these self maps.

Construction 2.11 concludes the second step in our second gambit for the prize!

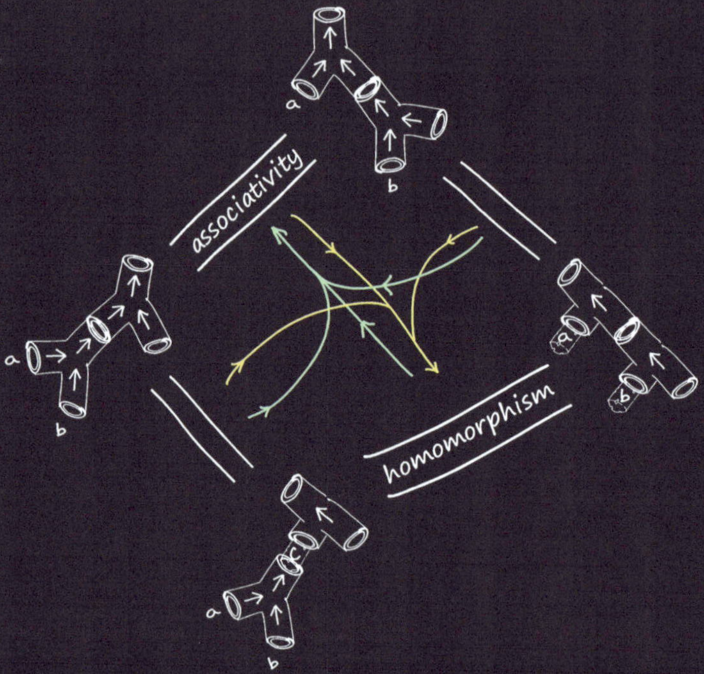

Construction 2.12 The above figure shows that given an associative magma M on S, we can canonically construct the **firefly** from M to a submagma of $\{S \to S\}$ under composition, and given such a firefly from a magma M' on S to a submagma of $\{S \to S\}$ under composition, we can show that M' is associative.

Construction 2.12 demonstrates that our second gambit succeeded: the subset of the set of magmas known as the set of associative magmas is therefore crowned as an object produced by a **metacanonical construction**: a canonical construction canonically selected because it lies entirely within the mirrored space of a mirror (correspondence).

Meaning 2.2 It follows from Construction 2.12 that associativity of a magma M on S means that its comet is a firefly from M to a submagma of $\{S \to S\}$ under composition.

2.3 Distributivity

Construction 2.13 *Given a magma M on S it follows from Result 2.4 that we can canonically construct a* **magma on** End(M) *by using composition as a binary operation on Construction 2.8. We will denote this magma as* (End$(M), \circ$)*. This is also trivially a canonical construction of a submagma of* $(\{S \to S\}, \circ)$.

Construction 2.14 *Given an associative magma on S we can use the image of the firefly from Construction 2.12 to canonically construct a* **submagma of** $(\{S \to S\}, \circ)$.

The similarity between Construction 2.14 and Construction 2.13 is striking. Both canonical constructions take a magma on S and produce a submagma of $(\{S \to S\}, \circ)$.

Result 2.5 Let $M = (S, \cdot)$ be the magma given to Construction 2.14, and let $M' = (S, \oplus)$ be the magma given to Construction 2.13. Then the submagma produced by Construction 2.14 is a subset of the submagma produced by Construction 2.13 precisely when the following equality holds for every $a, x, y \in S$:

$$a \cdot (x \oplus y) = (a \cdot x) \oplus (a \cdot y).$$

Requiring the magma M, in addition to being associative, to left-distribute over the magma M' allows us to use the image of the firefly from M to $(\{S \to S\}, \circ)$ as a canonical construction of a submagma of (End$(M'), \circ$).

Chapter 3

Groups

With a Surprise Appearance by a Category

3.1 Lasso

An inverse element of a in a group is usually described as an element d satisfying $d \cdot a = e = a \cdot d$. An inverse map of $f\colon X \to Y$ is usually described as a map $g\colon Y \to X$ satisfying $g \circ f = \mathrm{id}_X$ and $f \circ g = \mathrm{id}_Y$. Textbooks typically introduce these as separate notions. Yet each seems to carry a visible tattoo announcing the same parent: reverse direction, combine in both orders, and recover the appropriate identities.

Let us find a single parent definition from which both inverse elements and inverse maps arise as special cases. To do so we must first loosen the lasso.

Definition 3.1 Let $m\colon D \to S$ be a binary multiplication with set of labels $\mathcal{O}$ and typed sets $S_{X,Y}$ as in Machine 1.1. An element $\ell_X \in S_{X,X}$ is a **left identity** at $X \in \mathcal{O}$ if

$$m(\ell_X, a) = a$$

for every $Y \in \mathcal{O}$ and every $a \in S_{Y,X}$.

At a fixed label, a binary multiplication may have more than one left identity, or no left identity at all.

Definition 3.2 Let $m\colon D \to S$ be a binary multiplication with set of labels $\mathcal{O}$ and typed sets $S_{X,Y}$ as in Machine 1.1. An element $r_X \in S_{X,X}$ is a **right identity** at $X \in \mathcal{O}$ if

$$m(a, r_X) = a$$

for every $Y \in \mathcal{O}$ and every $a \in S_{X,Y}$.

At a fixed label, a binary multiplication may have more than one right identity, or no right identity at all.

Definition 3.3 An element $e_X \in S_{X,X}$ is an **identity** at the label X if it is simultaneously a left identity (Definition 3.1) and a right identity (Definition 3.2) at X.

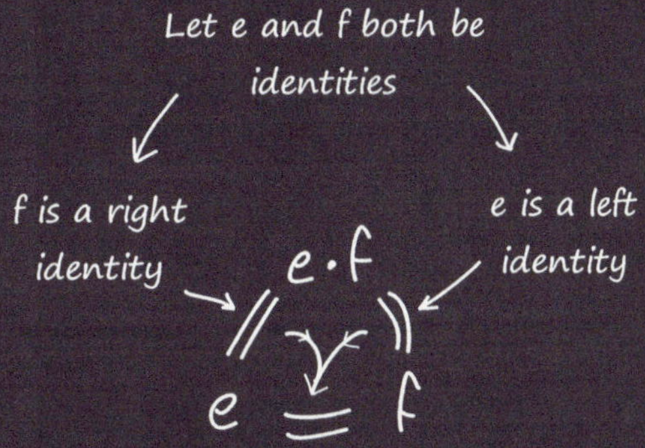

★ At each label X, a binary multiplication can have no more than one identity. If ℓ_X is a left identity and r_X is a right identity at X, then $m(\ell_X, r_X) = r_X$ by the left identity law and $m(\ell_X, r_X) = \ell_X$ by the right identity law. Thus they coincide and are the unique one-sided identities at X.

Definition 3.4 A binary multiplication is called **unital** if it has an identity at every label $X \in \mathcal{O}$. Its identities form the family $(e_X)_{X \in \mathcal{O}}$. Every $a \in S_{X,Y}$ then satisfies the left identity law $m(e_Y, a) = a$ and the right identity law $m(a, e_X) = a$.

Construction 3.1 Given a set $\mathcal{O}$ of sets, composition on the set of all maps between members of $\mathcal{O}$ is a canonical construction of a **unital binary multiplication**. For every $X \in \mathcal{O}$, the identity map id_X is its identity at X.

Definition 3.5 When the set of labels is a singleton, a left identity at its unique label (Definition 3.1) is called a **left identity element** e_L of the corresponding magma. Equivalently, $m(e_L, a) = a$ for every element a of the magma, so left multiplication by e_L is the identity map (see Construction 2.3 and Definition 1.1).

Definition 3.6 When the set of labels is a singleton, a right identity at its unique label (Definition 3.2) is called a **right identity element** e_R of the corresponding magma. Equivalently, $m(a, e_R) = a$ for every element a of the magma, so right multiplication by e_R is the identity map (see Construction 2.4 and Definition 1.1).

An element e of a magma is its **identity element** if it is simultaneously a left identity element (Definition 3.5) and a right identity element (Definition 3.6).

A binary operation that is unital as a binary multiplication (Definition 3.4)—equivalently, a magma that contains an identity element—is called a **unital magma**.

A semigroup (an associative magma) that contains the identity element is called a **monoid**.

Definition 3.7 Let $m\colon D \to S$ be a unital binary multiplication with set of labels $\mathcal{O}$, typed sets $S_{X,Y}$, and identity family $(e_Z)_{Z\in\mathcal{O}}$ as in Machine 1.1 and Definition 3.4. For $a \in S_{X,Y}$, an element $\ell \in S_{Y,X}$ is a **left inverse** of a under m if

$$m(\ell, a) = e_X.$$

Definition 3.8 Let $m\colon D \to S$ be a unital binary multiplication with set of labels $\mathcal{O}$, typed sets $S_{X,Y}$, and identity family $(e_Z)_{Z\in\mathcal{O}}$ as in Machine 1.1 and Definition 3.4. For $a \in S_{X,Y}$, an element $r \in S_{Y,X}$ is a **right inverse** of a under m if

$$m(a, r) = e_Y.$$

Definition 3.9 An element $d \in S_{Y,X}$ is an **inverse** of $a \in S_{X,Y}$ under a unital binary multiplication if it is simultaneously a left inverse (Definition 3.7) and a right inverse (Definition 3.8) of a. Equivalently, it satisfies both $m(d, a) = e_X$ and $m(a, d) = e_Y$.

Definition 3.10 Taking $\mathcal{O}$ to be a singleton specializes Definition 3.7, Definition 3.8, and Definition 3.9 to the following notions for a unital magma U with binary operation m and identity element e. A **left inverse** of $a \in U$ is an element $\ell \in U$ satisfying $m(\ell, a) = e$; a **right inverse** of a is an element $r \in U$ satisfying $m(a, r) = e$; and an **inverse element** of a (or simply an inverse of a) is an element $d \in U$ that is both a left inverse and a right inverse. Equivalently,

$$m(d, a) = e = m(a, d).$$

Definition 3.11 Specializing Definition 3.9 to the unital binary multiplication of composition in Construction 3.1, a map $g\colon Y \to X$ is an **inverse map** of a map $f\colon X \to Y$ if

$$g \circ f = \mathrm{id}_X,$$
$$f \circ g = \mathrm{id}_Y.$$

Definition 3.9 gives the common parent: **inverse under a unital binary multiplication**. Definition 3.10 is its one-label specialization to inverse elements, and Definition 3.11 is its specialization to inverse maps under composition.

Definition 3.12 An **associative binary multiplication** is a binary multiplication $m\colon D \to S$ that is associative as a binary map (Definition 2.1). Equivalently,

$$m(m(a, b), c) = m(a, m(b, c))$$

for all $a, b, c \in S$ whenever $\mathrm{dom}(a) = \mathrm{cod}(b)$ and $\mathrm{dom}(b) = \mathrm{cod}(c)$.

Definition 3.13 A **category** is a binary multiplication that is both unital (Definition 3.4) and associative (Definition 3.12). Its labels $X \in \mathcal{O}$ are called objects; an element $a \in S_{X,Y}$ is called an arrow from X to Y (also called a morphism); $m(a, b)$, usually written $a \circ b$, is called composition; and the identity e_X at X may also be written id_X and called the identity arrow at X.

Construction 3.2 For a set $\mathcal{O}$ of sets, associativity of composition makes the unital binary multiplication in Construction 3.1 the **category of all maps between members of** $\mathcal{O}$. Its objects are the sets in $\mathcal{O}$, and its arrows are the maps between them.

Definition 3.14 A category in which every arrow has an inverse in the sense of Definition 3.9 is called a **groupoid**.

Result 3.1 The arrows of a one-object category form a monoid under composition; conversely, every monoid canonically determines a one-object category.

Definition 3.15 A **group** is a groupoid (Definition 3.14) with exactly one object. Equivalently, by Result 3.1, it is a monoid containing an inverse for every element.

Trap 3.1 Even under associativity, an element may have several left inverses if it has no right inverse, or several right inverses if it has no left inverse. Without associativity, an element of a unital binary multiplication may have several of both kinds at once. Unlike the uniqueness of an identity, which follows from the identity laws alone, the uniqueness of an inverse requires associativity.

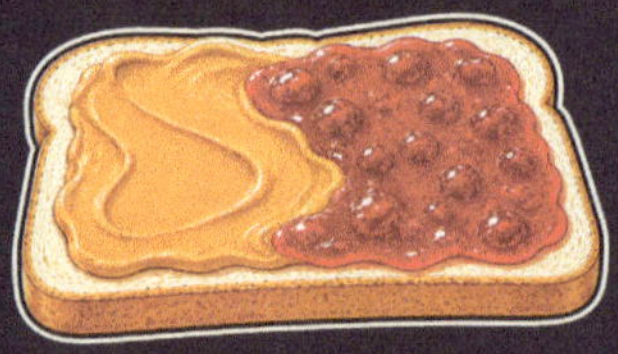

Associativity ensures that the inverse of each element in a group is unique. In this way the three properties of a group: identity, associativity, and a unique inverse for each element go together like peanut butter, jelly and toast!

Loosening to find the parent definition of an inverse meant that we were then tightening from a much wider lasso. This in turn led to us picking up the definition of a category when we once again tightened. Requiring every arrow to have an inverse gave a groupoid, and tightening to one object gave a group. We therefore have another canonically constructed subset of the set of all magmas on S: the set of groups on S, itself contained in the associative magmas constructed in Construction 2.7. We hope this completes the Lasso portion of our third gambit for the prize. Next we follow Arrow along the comet axis, hoping to canonically construct a subset of the comets on S which, when carried back across the correspondence in Construction 2.6, coincides with the groups just constructed.

3.2 Arrow

Construction 3.3 *Given a nonempty set S, we can canonically construct from the set of all permutations of S a* **magma under composition**, *denoted $(\mathrm{Sym}(S), \circ)$. This is then also trivially a canonical construction of a submagma of the magma constructed in Construction 2.10.*

Definition 3.16 *A firefly is colorfully called a* **dragon** *if, after restricting its codomain to its image, it is a magma isomorphism from the comet's magma onto a submagma of the magma of permutations in Construction 3.3.*

Construction 3.4 *Given a nonempty set S, we can use Definition 3.16 to canonically construct from the set of all comets on S in Construction 1.2 the* **subset consisting of the dragons**.

Construction 3.4 concludes the second step of our third gambit for the prize! Lasso constructed the groups on S, and Arrow has now constructed the dragons on S. It remains to determine whether the correspondence in Construction 2.6 carries the first subset exactly to the second.

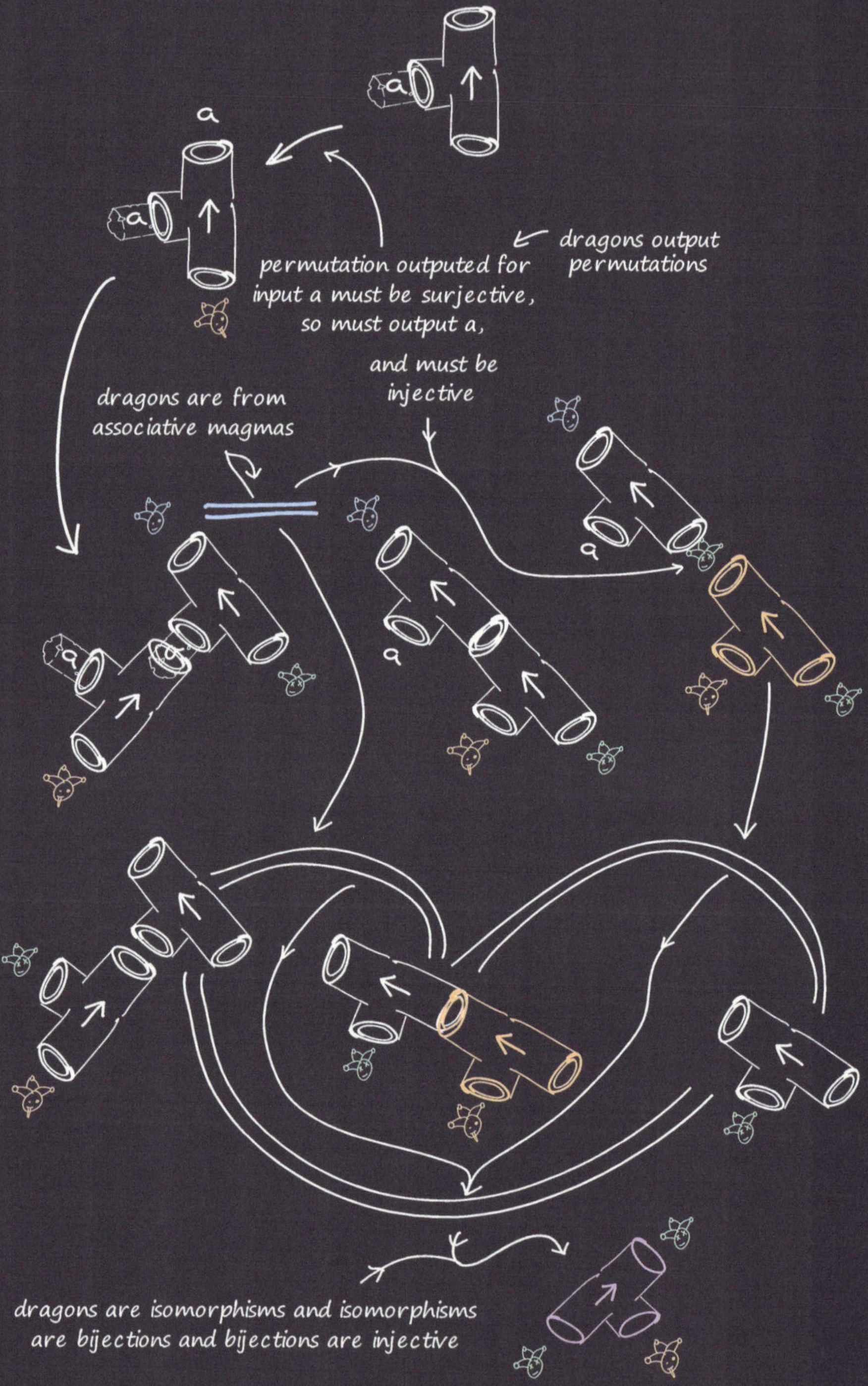

★ **Result 3.2** A dragon must be from a magma having a silly jester element such that, under the magma's operation, left and right multiplication by the silly jester are both the identity map.

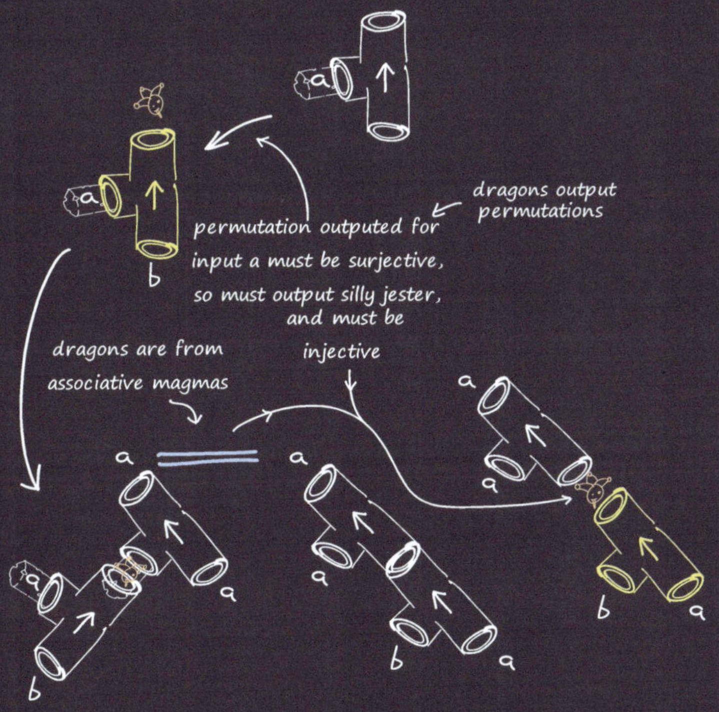

★ **Result 3.3** A dragon must be from a magma in which every element a has a corresponding element b such that multiplying a by b, in either order, yields the silly jester.

★ **Result 3.4** It is clear from Result 3.2 and Result 3.3 that a dragon must be from a magma that has an identity and an inverse for every element. From Construction 2.12 and the definition of a dragon as a firefly, this magma must also be associative. A dragon must therefore be from a group.

★ **Result 3.5** Under a group's firefly, an inverse element outputs an inverse map. Let b be an inverse of a. By Construction 2.12, composing the output for b with the output for a gives the output for ba, while composing them in the opposite order gives the output for ab. Both products are the identity element, whose output is the identity map because left multiplication by the identity element is the identity map. The output for b is therefore an inverse map of the output for a, in the sense of Definition 3.11.

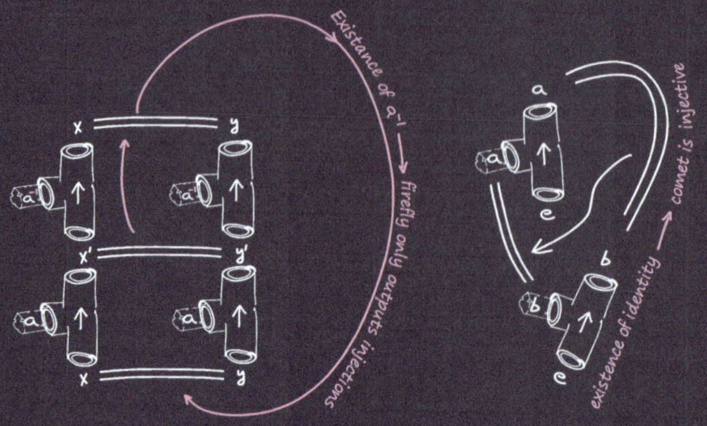

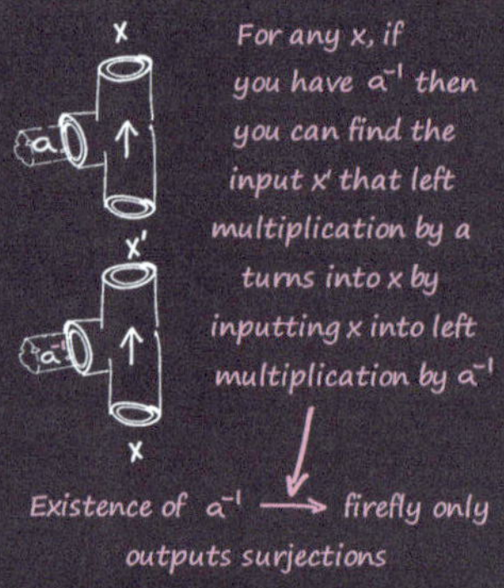

★ **Result 3.6** A map having an inverse map is injective because applying its inverse to equal outputs recovers equal inputs. It is surjective because every target is the output of the input obtained by applying the inverse map to that target. Thus every output of a group's firefly is a permutation. The figure also shows that a magma's comet is injective whenever the magma has an identity, so restricting the comet's codomain to its image makes the comet bijective. A firefly from a group must be a dragon (see Definition 3.16).

Together, Result 3.4 and Result 3.6 demonstrate that our third gambit succeeded: Lasso canonically constructed the set of all groups on S directly, while Arrow canonically constructed the dragons among the comets on S, which the correspondence in Construction 2.6 carries back to exactly that same subset of magmas. We therefore crown the set of all groups on S as an object produced by a **metacanonical construction**: a canonical construction canonically selected because it lies entirely within the mirrored space of a mirror (correspondence).

Meaning 3.1 Together, Result 3.6 and Result 3.4 show that, for a nonempty magma M, being a **group** means that its comet is a **dragon**.

Construction 3.5 Given a bijection $f: X \to Y$, we can canonically construct its **inverse map** $f^{-1}: Y \to X$ under composition. For every $y \in Y$, surjectivity of f guarantees the existence of an $x \in X$ such that $f(x) = y$, while injectivity of f guarantees that this x is unique. We can therefore define $f^{-1}(y) = x$. By construction, for every $x \in X$ and every $y \in Y$,

$$f^{-1}(f(x)) = x,$$
$$f(f^{-1}(y)) = y.$$

Thus f^{-1} is an inverse map of f in the sense of Definition 3.11.

Result 3.7 For a set S, the magma $(\mathsf{Sym}(S), \circ)$ constructed in Construction 3.3 is a group: composition is associative, id_S is its identity, and every permutation has an inverse permutation by Construction 3.5. It is called the **symmetric group** on S.

Definition 3.17 A subset H of a group G is a **subgroup** if the operation of G, restricted to H, makes H a group.

Result 3.8 The image of a group's firefly is a subgroup of the symmetric group on the group's underlying set. By Result 3.6, its elements are permutations. By Construction 2.12, composing two outputs gives the output for the product of their inputs, so the image is closed under composition. By Result 3.5, it contains the identity permutation and the inverse of each of its elements. Composition is associative by Result 2.1. The image is therefore a group under the restricted operation and hence a subgroup in the sense of Definition 3.17.

Result 3.9 **Cayley's theorem.** Every group is isomorphic to a subgroup of the symmetric group on its underlying set. By Result 3.6, the group's firefly, after restricting its codomain to its image, is a magma isomorphism from the original group onto that image. By Result 3.8, the image is a subgroup of the symmetric group.

Where next?

In this short text, we introduced the concept of metacanonical construction and found three examples: the sets of all magmas, semigroups, and groups on S. In all three cases, the metacanonical criterion was implemented using a mirror—and, in fact, the same mirror. What other familiar objects in algebra arise from mirror implementations of metacanonical construction? What other implementations are possible, and can they be composed with the filters developed here, either in series or in parallel? I hope this new perspective serves as a compass, guiding us toward what may be the most productive areas of algebra.

www.ingramcontent.com/pod-product-compliance
Lightning Source LLC
LaVergne TN
LVHW070627061126
851857LV00003B/30

* 9 7 9 8 9 9 7 8 3 3 8 0 0 *